This series aims at speedy, informal, and high level information on new developments in mathematical research and teaching. Considered for publication are:

1. Preliminary drafts of original papers and monographs

2. Special lectures on a new field, or a classical field from a new point of view

3. Seminar reports

4. Reports from meetings

Out of print manuscripts satisfying the above characterization may also be considered, if they continue to be in demand.

The timeliness of a manuscript is more important than its form, which may be unfinished and preliminary. In certain instances, therefore, proofs may only be outlined, or results may be presented which have been or will also be published elsewhere.

The publication of the *"Lecture Notes"* Series is intended as a service, in that a commercial publisher, Springer-Verlag, makes house publications of mathematical institutes available to mathematicians on an international scale. By advertising them in scientific journals, listing them in catalogs, further by copyrighting and by sending out review copies, an adequate documentation in scientific libraries is made possible.

Manuscripts
Since manuscripts will be reproduced photomechanically, they must be written in clean typewriting. Handwritten formulae are to be filled in with indelible black or red ink. Any corrections should be typed on a separate sheet in the same size and spacing as the manuscript. All corresponding numerals in the text and on the correction sheet should be marked in pencil. Springer-Verlag will then take care of inserting the corrections in their proper places. Should a manuscript or parts thereof have to be retyped, an appropriate indemnification will be paid to the author upon publication of his volume. The authors receive 25 free copies.

Manuscripts in English, German or French should be sent to Prof. Dr. A. Dold, Mathematisches Institut der Universität Heidelberg, Tiergartenstraße or Prof. Dr. B. Eckmann, Eidgenössische Technische Hochschule, Forschungsinstitut für Mathematik, Zürich.

Die *„Lecture Notes"* sollen rasch und informell, aber auf hohem Niveau, über neue Entwicklungen der mathematischen Forschung und Lehre berichten. Zur Veröffentlichung kommen:

1. Vorläufige Fassungen von Originalarbeiten und Monographien.

2. Spezielle Vorlesungen über ein neues Gebiet oder ein klassisches Gebiet in neuer Betrachtungsweise.

3. Seminarausarbeitungen.

4. Vorträge von Tagungen.

Ferner kommen auch ältere vergriffene spezielle Vorlesungen, Seminare und Berichte in Frage, wenn nach ihnen eine anhaltende Nachfrage besteht.

Die Beiträge dürfen im Interesse einer größeren Aktualität durchaus den Charakter des Unfertigen und Vorläufigen haben. Sie brauchen Beweise unter Umständen nur zu skizzieren und dürfen auch Ergebnisse enthalten, die in ähnlicher Form schon erschienen sind oder später erscheinen sollen.

Die Herausgabe der *„Lecture Notes"* Serie durch den Springer-Verlag stellt eine Dienstleistung an die mathematischen Institute dar, indem der Springer-Verlag für ausreichende Lagerhaltung sorgt und einen großen internationalen Kreis von Interessenten erfassen kann. Durch Anzeigen in Fachzeitschriften, Aufnahme in Kataloge und durch Anmeldung zum Copyright sowie durch die Versendung von Besprechungsexemplaren wird eine lückenlose Dokumentation in den wissenschaftlichen Bibliotheken ermöglicht.

Lecture Notes in Mathematics

A collection of informal reports and seminars
Edited by A. Dold, Heidelberg and B. Eckmann, Zürich

Series: Forschungsinstitut für Mathematik, ETH, Zürich · Adviser: K. Chandrasekharan

4

M. Arkowitz · C. R. Curjel

Dartmouth College, Hanover, N.H.
University of Washington, Seattle, Wash.
Forschungsinstitut für Mathematik,
Eidg. Techn. Hochschule, Zürich

Groups of Homotopy Classes

(Rank formulas and homotopy-commutativity)

1967

Springer-Verlag Berlin Heidelberg GmbH

Second revised edition.

ISBN 978-3-540-03900-6   ISBN 978-3-540-34968-6 (eBook)
DOI 10.1007/978-3-540-34968-6

All rights, especially that of translation into foreign languages, reserved. It is also forbidden to reproduce this book, either whole or in part, by photomechanical means (photostat, microfilm and/or microcard) or by other procedure without written permission from Springer-Verlag Berlin Heidelberg GmbH.

© by Springer-Verlag Berlin Heidelberg 1967.
Originally published by Springer-Verlag Berlin Heidelberg New York in 1967

Library of Congress Catalog Card Number 67-17948. Title No. 7324.

Contents

1. Introduction — 1
2. Groups of Finite Rank — 3
3. The Groups $[A,\Omega X]$ and their Homomorphisms — 10
4. Commutativity and Homotopy-Commutativity — 19
5. The Rank of the Group of Homotopy Equivalences — 26

Bibliography — 34

Acknowledgments

These notes were written while the authors were members of the Forschungsinstitut für Mathematik, Eidg. Technische Hochschule, Zürich, during the summer of 1964. We wish to thank Professor B. Eckmann for having provided us with an opportunity to work together at the Forschungsinstitut.

Some of the results reported here were obtained while the first named author was partially supported by Princeton University, and the second named author by the Institute for Advanced Study and Cornell University.

1. Introduction

Many of the sets that one encounters in homotopy classification problems have a natural group structure. Among these are the groups $[A,\Omega X]$ of homotopy classes of maps of a space A into a loop-space ΩX. Other examples are furnished by the groups $\mathcal{E}(Y)$ of homotopy classes of homotopy equivalences of a space Y with itself. The groups $[A,\Omega X]$ and $\mathcal{E}(Y)$ are not necessarily abelian. It is our purpose to study these groups using a numerical invariant which can be defined for any group. This invariant, called the rank of a group, is a generalisation of the rank of a finitely generated abelian group. It tells whether or not the groups considered are finite and serves to distinguish two infinite groups. We express the rank of subgroups of $[A,\Omega X]$ and of $\mathcal{E}(Y)$ in terms of rational homology and homotopy invariants. The formulas which we obtain enable us to compute the rank in a large number of concrete cases. As the main application we establish several results on commutativity and homotopy-commutativity of H-spaces.

Chapter 2 is purely algebraic. We recall the definition of the rank of a group and establish some of its properties. These facts, which may be found in the literature, are needed in later sections.

Chapter 3 deals with the groups $[A,\Omega X]$ and the homomorphisms $f^*: [B,\Omega X] \to [A,\Omega X]$ induced by maps $f: A \to B$. We prove a general theorem on the rank of the intersection of coincidence subgroups (Theorem 3.3). As consequences we obtain formulas for the rank of the kernel of f^*, the image of f^* and, in the case $A = B$, of the fixed point set of a collection of induced endomorphisms. These expressions involve $f^*: H^*(B) \to H^*(A)$ and the rank of the homotopy groups of ΩX. In the remainder of the chapter we consider several corollaries of these results.

In Chapter 4 we apply the theorems of Chapter 3 to study homotopy-commutative maps of a topological group G. An element of $[G \times G, G]$ is called a commutative product if it factors through the symmetric square of G, and a homotopy-commutative product if it is invariant under the switching map $G \times G \to G \times G$. We prove in Theorem 4.5: (i) There exist nontrivial commutative and homotopy-commutative products on G. (ii) A suitable multiple of a homotopy-commutative product is commutative. (iii) G admits an infinite number of different commutative and homotopy-commutative products if and only if the Betti numbers of G satisfy an easily verifiable relation. Actually we work in a more general setting and consider commutative and homotopy-commutativity elements of $[A \times A, \Omega Y]$. The chapter concludes with various remarks in which we elucidate the relation of our work to that of other authors, mention further results and discuss a general conjecture on homotopy-associative H-spaces.

In the last chapter we determine the rank of $\mathcal{E}(Y)$ if Y is a 1-connected space whose rational cohomology is an exterior algebra on odd-dimensional generators.

Throughout these notes we only consider based spaces of the homotopy type of a connected CW-complex with finitely generated homology and homotopy groups in each dimension. All maps and homotopies are to preserve base points. We always identify the cohomology group $H^n(A;\pi)$ with $[A,K(\pi,n)]$.

We are grateful to P. J. Hilton for having pointed out to us an error in the first edition of these notes.

2. Groups of finite rank

We use the following notation and terminology. All groups are written additively. A group is said to be <u>periodic</u> if all its elements are of finite order. As usual, Z stands for the additive group of the integers. The kernel (image) of a homomorphism f is denoted by Ker f (Im f). If H is a normal subgroup of G we write $H \triangleleft G$.

It was K.A. Hirsch who first defined an invariant for solvable groups satisfying the maximal condition for subgroups* which generalizes the notion of the rank of a finitely generated abelian group [1938]. Jennings called this invariant "rank" and discussed some of its properties [1955]. The following (slightly more general) definition of a group of finite rank is due to Zassenhaus [1958, p. 241].

<u>Definition 2.1.</u> A group G is said to be of <u>finite rank</u>, written $\rho(G) < \infty$, if there exists a chain of subgroups
$$G = G_0 \triangleright G_1 \triangleright \ldots \triangleright G_t = 0$$
such that G_i/G_{i+1} is either infinite cyclic or periodic, $i = 0, \ldots t-1$. Such a chain will be called a ρ-chain of G.

---

* A group G is <u>solvable</u> if there exists a chain of subgroups
  $G = G_0 \triangleright G_1 \triangleright \ldots \triangleright G_i \triangleright \ldots \triangleright G_t = 0$ such that all factors G_i/G_{i+1} are abelian; G satisfies the <u>maximal condition for subgroups</u> if every ascending chain of distinct subgroups is finite, or, equivalently, if G and all its subgroups are finitely generated.

The following Proposition is just a rewording of Theorem 1.42 of [K.A. Hirsch, 1938].

Proposition 2.2. Let G be a group of finite rank. The number of infinite cyclic factors of any ρ-chain of G is an invariant of G.

Proof: a) The following fact will be used repeatedly in the proof of the Proposition: Let A, A' be subgroups of B such that

$$A \triangleleft B, \quad A \triangleleft A' \triangleleft B.$$

If B/A is periodic, so are B/A' and A'/A; if $B/A \cong Z$ then one of the two groups B/A', A'/A is infinite cyclic and the other periodic. This is seen as follows. The inclusion $A' \subseteq B$ gives rise to a monomorphism $\Psi: A'/A \to B/A$ whose image is a normal subgroup of B/A. We identify Im Ψ with A'/A and obtain the isomorphism

$$B/A' \cong \frac{B/A}{A'/A}.$$

If B/A is periodic, so are obviously B/A' and A'/A; if $B/A \cong Z$ and $A'/A \cong 0$ then $B/A' \cong Z$; if $B/A \cong Z$ and $A'/A \neq 0$ then $A'/A \cong Z$ and $B/A' \cong Z/mZ = Z_m$ for some integer m.

b) Now we prove the Proposition. Let $\{G_i\}$ and $\{G'_j\}$ be two ρ-chains of G. By the Jordan-Hölder-Schreier refinement theorem these two ρ-chains can be refined to chains $\{_1G_k\}$ and $\{_1G'_\ell\}$ with the same factors up to a permutation. Let $H_1, \ldots, H_r$ be the groups inserted between G_i and G_{i+1} for the refinement of $\{G_i\}$:

$$(*) \quad G_{i+1} \triangleleft H_1 \triangleleft H_2 \triangleleft \ldots \triangleleft H_r \triangleleft G_i.$$

(1) If G_i/G_{i+1} is periodic, so are G_i/H_r and H_r/G_{i+1} by a) above, and similarly H_n/H_{n-1} and H_{n-1}/G_{i+1} are in turn periodic for $n = r, r-1, \ldots, 2$. Thus all factors of $(*)$ are periodic if G_i/G_{i+1} is periodic.

(2) If $G_i/G_{i+1} \cong Z$ then a repeated application of the remark in a) shows that exactly one of the factors of $(*)$ is infinite cyclic and the rest periodic.

It follows from (1) and (2) that the number of infinite cyclic factors remains unchanged if $\{G_i\}$ is refined to $\{_1G_k\}$ and $\{G'_j\}$ to $\{_1G'_\ell\}$. Since $\{_1G_k\}$ and $\{_1G'_\ell\}$ have the same factors up to a permutation, the number of infinite cyclic factors of a ρ-chain of G is an invariant of G.

In view of this Proposition the following definition makes sense.

Definition 2.3. If G is a group of finite rank then the number $\rho(G)$ of infinite cyclic factors of any ρ-chain of G is called the <u>rank</u> of G.

It is clear that a finitely generated abelian group has finite rank, and that the rank of such a group is just the rank in the usual sense. Most of the results of this section are simply generalisations of properties of the rank of an abelian group.

Next we introduce notions analogous to those of C-theory of abelian groups.

Definition 2.4. Let f: A → B be a homomorphism of groups. We call f an <u>F-monomorphism</u> if Ker f is periodic, and we call f an <u>F-epimorphism</u> if for any b ∈ B there exists an integer n ≠ 0 such that nb ∈ Im f. By an <u>F-isomorphism</u> is meant a map which is both an F-monomorphism and F-epimorphism.

Most of the assertions of the following Proposition can be found in the previously cited literature.

Proposition 2.5 (properties of rank).

(a) Let G be a group of finite rank. Then $\rho(G) = 0$ if and only if G is periodic.

(b) Let H be a subgroup of a group G of finite rank. Then $\rho(H) \leq \rho(G)$, and $\rho(H) = \rho(G)$ if the inclusion $H \subseteq G$ is an F-epimorphism. Furthermore $\rho(G) = \rho(H)$ implies that the inclusion $H \subseteq G$ is an F-epimorphism provided every periodic subquotient (i.e., quotient of a subgroup) of G is finite.

(c) Let N be a normal subgroup of G. Then G is of finite rank if and only if both G/N and N are of finite rank. In this case

$$\rho(G) = \rho(G/N) + \rho(N).$$

Proof. (a) If G is periodic then obviously $\rho(G) = 0$. If on the other hand $\rho(G) = 0$ then all factors of a ρ-chain

$$G = G_0 \triangleright G_1 \triangleright \ldots \triangleright G_t = 0$$

are periodic. Since the extension of a periodic group by a periodic group is periodic, all G_i are in turn periodic. Hence G is periodic.

(b) Let $\{G_i\}$ be a ρ-chain of G. Clearly $G_{i+1} \triangleleft G_i$ implies $H \cap G_{i+1} \triangleleft H \cap G_i$. The inclusion $H \cap G_i \subseteq G_i$ induces a monomorphism ϕ:

$$\begin{array}{ccccccccc}
0 & \to & G_{i+1} & \to & G_i & \to & G_i/G_{i+1} & \to & 0 \\
& & \uparrow & & \uparrow & & \uparrow \phi & & \\
0 & \to & H \cap G_{i+1} & \to & H \cap G_i & \to & H \cap G_i / H \cap G_{i+1} & \to & 0.
\end{array}$$

Thus the factors of the chain $\{H \cap G_i\}$ are subgroups of the factors of $\{G_i\}$, and therefore are periodic or infinite cyclic. Hence $\rho(H) \leq \rho(G)$. Now let the inclusion $H \subseteq G$ be an F-epimorphism. Then the inclusion $H \cap G_i \subseteq G_i$ is also an F-epimorphism. It now follows from the diagram above that ϕ is an F-isomorphism. Hence $G_i/G_{i+1} \cong Z$ implies $H \cap G_i / H \cap G_{i+1} \cong Z$. Therefore the number of infinite cyclic factors of the ρ-chain $\{H \cap G_i\}$ is the same as that of the ρ-chain $\{G_i\}$. Thus $\rho(H) = \rho(G)$.

Now we come to the last assertion of (b). Let $[G:H]$ denote the index of H in G. Observe that $[G:H] < \infty$ implies that $H \subseteq G$ is an F-epimorphism: If $mg \notin H$ for all integers $m \neq 0$ then the cosets $mg + H$ are all distinct. Thus it suffices to prove that $[G:H] < \infty$. Consider a ρ-chain $\{G_i\}$ of G and $\{G_i \cap H\}$ of H as in the diagram above. Since $\rho(G) = \rho(H)$ we see that G_i/G_{i+1} is finite or infinite cyclic precisely when the same holds for $H \cap G_i/H \cap G_{i+1}$. Therefore $[G_i/G_{i+1}: H \cap G_i/H \cap G_{i+1}] < \infty$ for all i. The proof of $[G:H] < \infty$ now proceeds by induction on the rank of G. If $\rho(G) = 0$, then G is finite, and $[G:H] < \infty$. Assume the assertion true for groups of rank $\leq n - 1$ and let $\rho(G) = \rho(H) = n$. Choose i in the preceding diagram such that $\rho(G_{i+1}) = n - 1$, $\rho(G_i) = n$. By the inductive hypothesis $[G_{i+1}: H \cap G_{i+1}] < \infty$, and from above we have $[G_i/G_{i+1}: H \cap G_i/H \cap G_{i+1}] < \infty$. Now apply the following general fact about groups: Let $A_1 \triangleleft A$, $B \subseteq A$, $B_1 = B \cap A_1$. Then $[A:B] = [A_1:B_1][A/A_1:B/B_1]$. Thus $[G_i: H \cap G_i] < \infty$. By again applying the same remark to $[G_k: H \cap G_k]$ for $k < i$ we eventually obtain $[G:H] < \infty$.

(c) Assume $\rho(G) < \infty$. Then $\rho(N) < \infty$ by (b). The projection $\psi: G \to G/N$ applied to a ρ-chain of G gives a ρ-chain of G/N. Thus $\rho(G/N) < \infty$. On the other hand suppose $\rho(G/N) < \infty$ and $\rho(N) < \infty$. Let $\{H_i\}$ be a ρ-chain of G/N, $i = 1,\ldots,s$, and $\{N_j\}$ a ρ-chain of N, $j = 1,\ldots t$. Then

$$G \triangleright \psi^{-1}(H_1) \triangleright \ldots \triangleright \psi^{-1}(H_s) = \psi^{-1}(0) = N \triangleright N_1 \triangleright \ldots \triangleright N_t = 0$$

is a ρ-chain of G, and clearly $\rho(G) = \rho(G/N) + \rho(N)$.

Following the terminology of [K.A. Hirsch, 1938] we define an
<u>S-group</u> to be a solvable group satisfying the maximal condition
for subgroups. We restate a result of K.A. Hirsch the first part of
which shows that the class of groups of finite rank is considerably
larger than the class of finitely generated abelian groups.

<u>Lemma 2.6.</u> (a) Any S-group is of finite rank. (b) Let G be an
S-group. Then $\rho(G) = 0$ if and only if G is finite.

<u>Proof.</u>   (a) Let G be an S-group and consider a chain

(*) $G = G_0 \rhd G_1 \rhd \ldots \rhd G_t = 0$

with abelian factors. Since G satisfies the maximal condition for
subgroups all factors of (*) are of the form $F + T$, where F is a
finitely generated free abelian group and T a finite abelian group.
In order to refine (*) to a ρ-chain we apply repeatedly the following
elementary remark: Let

$$0 \to H' \to H \xrightarrow{\phi} A + B \to 0$$

be exact, with $A + B$ abelian. Write $K = \phi^{-1}(A)$. Then $H' \lhd K \lhd H$,
$K/H' \cong A$ and $H/K \cong B$.

(b) In view of Proposition 2.5(a) it suffices to show that
any periodic S-group is finite. Let G be a periodic S-group and consider
the chain (*) of (a). Since G is periodic all factors of (*) are finite
abelian groups. The extension of a finite group by a finite group is
finite. Thus G is a finite group.

<u>Remarks 2.7.</u> (a) There exist groups of any given finite rank which
are <u>not</u> S-groups. Let $\mathcal{Z}_m$ be the direct sum of infinitely many copies
of Z_m, and let K be any finite group which is not solvable. Denote by
F_n the free abelian group on n letters. Then $\rho(F_n + \mathcal{Z}_m + K) = n$,
but $F_n + \mathcal{Z}_m + K$ is not solvable and does not satisfy the maximal
condition for subgroups.

(b) Any finitely generated nilpotent group is of finite rank because (i) any nilpotent group is solvable, and (ii) any finitely generated nilpotent group satisfies the maximal condition for subgroups (see, e.g., M. Hall [1959, p. 153]).

(c) A nonabelian free group F on two letters x and y is <u>not</u> of finite rank. For let (x) be the infinite cyclic group generated by x. By deleting y from each word of F, one defines an epimorphism of F onto $(x) \cong Z$ with kernel ϕ_1. Since (x) is abelian ϕ_1 contains the commutator subgroup of F. Hence ϕ_1 is a nonabelian free group. Now one picks a generator x_1 of ϕ_1 and repeats the argument to define a nonabelian free group ϕ_2 such that $\phi_1/\phi_2 \cong Z$. Thus we obtain an <u>infinite</u> sequence

$$F \triangleright \phi_1 \triangleright \phi_2 \triangleright \ldots \triangleright \phi_n \triangleright \ldots$$

such that $\phi_i/\phi_{i+1} \cong Z$. Therefore F is not of finite rank.

It is the following Lemma on which many of the arguments of Section 3 hinge.

<u>Lemma 2.8.</u> Let A,B,C be groups of finite rank and j,q homomorphisms

$$A \xrightarrow{j} B \xrightarrow{q} C$$

with the following properties: j is an F-monomorphism, q is an F-epimorphism, qj = 0 and the inclusion Im j ⊆ Ker q is an F-epimorphism. Then

$$\rho(B) = \rho(A) + \rho(C).$$

<u>Proof.</u> Consider the exact sequences

(a) 0 → Ker j → A → Im j → 0
(b) 0 → Ker q → B → Im q → 0.

Then $\rho(\text{Ker } j) = 0$ because j is an F-monomorphism. To say that q is an F-epimorphism means that the inclusion Im q ⊆ C is an F-epimorphism. Thus $\rho(\text{Im } q) = \rho(C)$ by Proposition 2.5(b), and we obtain from the sequences (a),(b) the relations

(a') $\rho(A) = \rho(\text{Im } j)$
(b') $\rho(B) = \rho(\text{Ker } q) + \rho(C)$.

Since the inclusion Im j ⊆ Ker q is an F-epimorphism, ρ(Im j) = ρ(Ker q) again by Proposition 2.5(b). This latter relation and (a'),(b') imply ρ(B) = ρ(A) + ρ(C).

3. The Groups $[A,\Omega X]$ and Their Homomorphisms

In this section we consider the group $[A,\Omega X]$ of homotopy classes of maps of the space A into the loop space ΩX and the homomorphisms $f^*: [B,\Omega X] \to [A,\Omega X]$ induced by a continuous map $f: A \to B$. We deal only with spaces with base points and maps and homotopies that preserve base points. Furthermore, it is assumed that all spaces have the homotopy type of connected CW-complexes with finitely generated homology and homotopy groups in each dimension.

Our main objective in this section is to find computable expressions for the rank of the kernel of f^*, of the image of f^* (Proposition 3.5) and, in the case A = B, of the fixed point set of a collection of induced endomorphisms (Proposition 3.9). These results and various others follow immediately from the main theorem (Theorem 3.3).

<u>We shall always assume in this section that the space X is simply connected and has only a finite number of nontrivial homotopy groups.</u> This is not a restriction in the common situation when A and B are finite dimensional CW-complexes. For in this case the following lemma shows that it suffices to consider Postnikov sections.

Lemma 3.1. Let A be an N-dimensional CW-complex and $\phi: Z \to Z'$ an N-equivalence. Then

$$\phi_*: [A,Z] \to [A,Z']$$

is a bijection. In particular, if Y is a simply connected space with a Postnikov decomposition $\{Y^n, k^n\}$, then the projection of Y onto Y^{N+1} induces an isomorphism

$$[A, \Omega Y] \cong [A, \Omega Y^{N+1}].$$

This lemma is an immediate consequence of Theorem 2 of [J.H.C. Whitehead, 1949].

The following theorem is the main result of this section.

Theorem 3.3. Let $f_i, g_i: A \to B$ be a collection of maps, $i \in I$, which induce homomorphisms $f_i^*, g_i^*: [B, \Omega X] \to [A, \Omega X]$ and $f_i^n, g_i^n: H^n(B) \to H^n(A)$. Denote by $C(f_i^*, g_i^*) \subseteq [B, \Omega X]$ the subgroup of all $\alpha \in [B, \Omega X]$ such that $f_i^*(\alpha) = g_i^*(\alpha)$, and define $C(f_i^n, g_i^n) \subseteq H^n(B)$ similarly. Then

$$\rho(\bigcap_{i \in I} C(f_i^*, g_i^*)) = \sum_m \rho(\bigcap_{i \in I} C(f_i^m, g_i^m)) \cdot \rho(\pi_m(\Omega X)).$$

Proof. Let $\{\Omega X^n, \Omega k^n\}$ be a Postnikov decomposition of ΩX obtained by applying the loop functor to a Postnikov decomposition $\{X^n, k^n\}$ of X. Consider one of the fibrations, say $\Omega X^{n+1} \to \Omega X^n$, with fibre $\Omega^2 K = K(\pi_n(\Omega X), n)$. Then there is a commutative diagram

$$\begin{array}{ccccc}
[B, \Omega^2 K] & \xrightarrow{j} & [B, \Omega X^{n+1}] & \xrightarrow{\phi} & [B, \Omega X^n] \\
\downarrow {f_i^*, g_i^*} & & \downarrow {f_i^*, g_i^*} & & \downarrow {f_i^*, g_i^*} \\
[A, \Omega^2 K] & \xrightarrow{j'} & [A, \Omega X^{n+1}] & \xrightarrow{\phi'} & [A, \Omega X^n],
\end{array}$$

where the horizontal arrows are induced either by the inclusion $\Omega^2 K \to \Omega X^{n+1}$ or the projection $\Omega X^{n+1} \to \Omega X^n$. We set $C' = \bigcap_{i \in I} C(f_i^*, g_i^*)$ $\subseteq [B, \Omega^2 K]$, and $C \subseteq [B, \Omega X^{n+1}]$ and $C'' \subseteq [B, \Omega X^n]$ are similarly defined. Then j and ϕ induce homomorphisms j'' and ϕ'' in the sequence

$$(*) \qquad C' \xrightarrow{j''} C \xrightarrow{\phi''} C''.$$

We show that this sequence satisfies the hypothesis of Lemma 2.8.

Consider the exact sequence

$$\to [B, \Omega^2 X^n] \xrightarrow{\Omega^2 k^n} [B, \Omega^2 K] \xrightarrow{j} [B, \Omega X^{n+1}] \xrightarrow{\phi} [B, \Omega X^n] \xrightarrow{\Omega k^n} [B, \Omega K].$$

By a result of Thom [1956] all Postnikov invariants of ΩX are of finite order. Therefore j and j' are F-monomorphisms (see Definition 2.4). Hence j'' in the sequence (*) is an F-monomorphism.

Now let $\alpha \in \operatorname{Ker} \phi''$. Then $\alpha = j(\beta)$ for some $\beta \in [B, \Omega^2 K]$. But $j'(f_i^* - g_i^*)(\beta) = (f_i^* - g_i^*)j(\beta) = (f_i^* - g_i^*)(\alpha) = 0$. Since j' is an F-monomorphism, there is an integer $M > 0$ such that $(f_i^* - g_i^*)(M\beta) = 0$ for all $i \in I$. Therefore $M\alpha = j''(M\beta)$, i.e., the inclusion $\operatorname{Im} j'' \subseteq \operatorname{Ker} \phi''$ is an F-epimorphism.

Next we show that ϕ'' is an F-epimorphism. Since Ωk^n has finite order there exists an integer N such that the N-fold sum $\operatorname{Id} + \ldots + \operatorname{Id}$ of the identity map of ΩX^n can be lifted to a map $\nu: \Omega X^n \to \Omega X^{n+1}$. Then $\alpha \in C''$ implies $\nu \alpha \in C$ and $\phi''(\nu \alpha) = N\alpha$:

$$A \xrightarrow{f_i, g_i} B \xrightarrow{\alpha} \Omega X^n \xrightarrow{\operatorname{Id}+\ldots+\operatorname{Id}} \Omega X^n$$

with $\nu: \Omega X^n \to \Omega X^{n+1}$ and projection $\Omega X^{n+1} \to \Omega X^n$.

Therefore ϕ'' is an F-epimorphism.

To summarize: In the sequence (*), j'' is an F-monomorphism, and ϕ'' and the inclusion $\operatorname{Im} j'' \subseteq \operatorname{Ker} \phi''$ are F-epimorphisms.

Furthermore we mention a general homological fact. Let π, π' be abelian groups and let $f^\pi: H^m(B;\pi) \to H^m(A;\pi)$ be the cohomology homomorphism induced by $f: A \to B$. Then

$$\bigcap_{i \in I} C(f_i^{\pi+\pi'}, g_i^{\pi+\pi'}) \cong \bigcap_{i \in I} C(f_i^\pi, g_i^\pi) + \bigcap_{i \in I} C(f_i^{\pi'}, g_i^{\pi'}).$$

By writing a finitely generated abelian group π as the direct sum of infinite cyclic groups and a periodic group, it follows from the preceding isomorphism that

$$(**) \qquad \rho(\bigcap_{i \in I} C(f_i^\pi, g_i^\pi)) = \rho(\bigcap_{i \in I} C(f_i^m, g_i^m)) \cdot \rho(\pi),$$

where $f_i^m = f_i^Z$, $g_i^m = g_i^Z : H^m(B) \to H^m(A)$.

Now we prove Theorem 3.3 by induction on the stages of a Postnikov decomposition of ΩX. By $(**)$, the theorem holds for the first stage. With the above notation let us assume for the induction that

$$\rho(C'') = \sum_{m < n} \rho(\bigcap_{i \in I} C(f_i^m, g_i^m)) \cdot \rho(\pi_m(\Omega X^n)).$$

Lemma 2.8 applied to the sequence $(*)$ yields

$$\rho(C) = \rho(C') + \rho(C'').$$

In view of the equality $\rho(C') = \rho(\bigcap_{i \in I} C(f_i^n, g_i^n)) \cdot \rho(\pi_n(\Omega X))$

obtained from $(**)$ we have

$$\rho(C) = \sum_{m < n+1} \rho(\bigcap_{i \in I} C(f_i^m, g_i^m)) \cdot \rho(\pi_m(\Omega X^{n+1})).$$

This completes the induction. Since X has only a finite number of homotopy groups, the theorem is proved.

The rest of the results in this section are all simple consequences of the preceding theorem. A very special case of the theorem occurs by setting $f_i = g_i = 0$ (the constant map) for all i in any index set I.

Thus we obtain

Corollary 3.4 [Arkowitz-Curjel, 1963_{II}]. $\rho([A,\Omega X]) = \sum_m \beta_m(A)\rho(\pi_m(\Omega X))$ where $\beta_m(A)$ is the m-th Betti number of A.

Another immediate consequence of Theorem 3.3 is

Proposition 3.5. Let $f: A \to B$ induce $f^*: [B,\Omega X] \to [A,\Omega X]$ and $f^n: H^n(B) \to H^n(A)$. Then

(a) $\rho(\text{Ker } f^*) = \sum_m \rho(\text{Ker } f^m) \cdot \rho(\pi_m(\Omega X))$;

(b) $\rho(\text{Im } f^*) = \sum_m \rho(\text{Im } f^m) \cdot \rho(\pi_m(\Omega X))$.

Proof. The proof of (a) follows from Theorem 3.3 by setting $f_i = f$ and $g_i = 0$ for all i in any index set I.

To prove (b) we observe that $\rho([B,\Omega X]) = \rho(\text{Ker } f^*) + (\text{Im } f^*)$ by Proposition 2.5. Since $\rho([B,\Omega X]) = \sum_m \beta_m(B)\rho(\pi_m(\Omega X))$ by Corollary 3.4, we obtain

$$\rho(\text{Im } f^*) = \sum_m \beta_m(B)\rho(\pi_m(\Omega X)) - \sum_m \rho(\text{Ker } f^m)\rho(\pi_m(\Omega X))$$

$$= \sum_m (\beta_m(B) - \rho(\text{Ker } f^m))\rho(\pi_m(\Omega X))$$

$$= \sum_m \rho(\text{Im } f^m)\rho(\pi_m(\Omega X)),$$

and the proof of the Proposition is complete.

Proposition 3.5 enables us to give a short proof of Proposition 4 of [Arkowitz-Curjel, 1963_I] :

Corollary 3.6. Let $g: A \to \Omega X$ and let Q denote the group of rationals. If $g^* = 0 : H^n(\Omega X; Q) \to H^n(A; Q)$ for all n for which $\rho(\pi_n(\Omega X)) \neq 0$, then the homotopy class of g is an element of finite order in $[A,\Omega X]$.

Proof. It follows from our hypothesis that $\rho(\text{Im } g^n) = 0$, where $g^n\colon H^n(\Omega X) \to H^n(A)$. Thus the image of the homomorphism $g^*\colon [\Omega X, \Omega X] \to [A, \Omega X]$ is a periodic group by Proposition 3.5(b) and Proposition 2.5(a). Since the homotopy class of g equals $g^*(\text{Id})$, it is of finite order.

Corollary 3.7. In the notation of Proposition 3.5

    (a) Ker f* is finite if and only if $\rho(\text{Ker } f^m)\rho(\pi_m(\Omega X)) = 0$ for all m.

    (b) Im f* is finite if and only if $\rho(\text{Im } f^m)\rho(\pi_m(\Omega X)) = 0$ for all m.

Proof. We first show that $[A, \Omega X]$ is an S-group (see the discussion before Lemma 2.6). As in the proof of Theorem 3.3 we consider the exact sequence obtained from a Postnikov decomposition of X:

$$[A, K(\pi_n(\Omega X), n)] \to [A, \Omega X^{n+1}] \to [A, \Omega X^n].$$

Since $[A, K(\pi, r)] \cong H^r(A; \pi)$ is an S-group, we may assume for induction that $[A, \Omega X^n]$ is an S-group. But subgroups, quotients and extensions of S-groups are again S-groups, and so $[A, \Omega X^{n+1}]$ is an S-group. Thus $[A, \Omega X]$ and, consequently, Ker f* and Im f* are all S-groups. The Corollary now follows from Proposition 3.5 and Lemma 2.6(b).

Corollary 3.8. Let the induced homomorphism $f^*\colon H^*(B; Q) \to H^*(A; Q)$ be a monomorphism (resp., epimorphism), where Q denotes the group of rationals. Then $f^*\colon [B, \Omega X] \to [A, \Omega X]$ is an F-monomorphism (resp., F-epimorphism).

For the proof that f* is an F-epimorphism one needs in addition to Proposition 3.5 the second part of Proposition 2.5 (b) and the fact that $[A, \Omega X]$ is an S-group.

We now turn to a consequence of Theorem 3.3 dealing with the elements of $[A,\Omega X]$ which are fixed points of a collection of induced endomorphisms. Our interest in fixed points stems from the fact a multiplication of a topological group G is homotopy-commutative precisely when it is a fixed point of the endomorphism of $[G \times G;G]$ which is induced by the switching map $G \times G \to G \times G$. In the general situation, let Λ be an arbitrary subset of $[A,A]$. We denote by $[A,\Omega X]^\Lambda$ the subgroup of $[A,\Omega X]$ consisting of all α such that $f^*(\alpha) = \alpha$ for all $f \in \Lambda$. We similarly define $H^m(A;\pi)^\Lambda$. Then the following Proposition is obtained from Theorem 3.3 by setting $\Lambda = \{f_i \mid i \in I\}$ and $g_i = \text{Id}$ for all i.

Proposition 3.9. If Λ is any subset of $[A,A]$ then

$$\rho([A,\Omega X]^\Lambda) = \sum_m \rho(H^m(A)^\Lambda)\rho(\pi_m(\Omega X)).$$

For a special case of Proposition 3.9 we consider a group Γ of homeomorphisms of A. Though it is not Γ but the image of Γ in $[A,A]$ which operates on $[A,\Omega X]$ we retain the notation $[A,\Omega X]^\Gamma$ for the subgroup of invariant elements. We denote by A/Γ the orbit space of A under the action of Γ and by $q: A \to A/\Gamma$ the projection.

Any composite map $A \xrightarrow{q} A/\Gamma \to \Omega X$ clearly determines an element of $[A,\Omega X]^\Gamma$. Conversely, one may ask: Which elements of $[A,\Omega X]^\Gamma$ can be factored through A/Γ? The following Proposition which generalizes a corollary of a result of Conner and Grothendieck (see [Borel, 1960, p. 38]) gives conditions under which an appropriate multiple of any element of $[A,\Omega X]^\Gamma$ can be factored through A/Γ.

Proposition 3.10. Let Γ be a finite group of homeomorphisms of the finite CW-complex A, and let $q^*: [A/\Gamma,\Omega X] \to [A,\Omega X]$ be the homomorphism induced by the projection p. Then

$$\rho([A/\Gamma,\Omega X]) = \rho(\text{Im } q^*) = \rho([A,\Omega X]^\Gamma).$$

Proof. The hypotheses on A and Γ imply that the n-th integer Čech cohomology group of A/Γ can be identified with $[A/\Gamma,K(Z,n)]$. Therefore III, Corollary 2.3 of [Borel, 1960] yields

$$(*) \quad \rho([A/\Gamma,K(Z,m)]) = \rho(\text{Im } q^m) = \rho([A,K(Z,m)]^\Gamma),$$

where $q^m: [A/\Gamma,K(Z,m)] \to [A,K(Z,m)]$ is induced by q. Now we multiply each term of (*) with $\rho(\pi_m(\Omega X))$ and sum over m. In view of Corollary 3.4, Proposition 3.5(b) and Proposition 3.8 we obtain

$$\rho([A/\Gamma,\Omega X]) = \rho(\text{Im } q^*) = \rho([A,\Omega X]^\Gamma),$$

where $q^*: [A/\Gamma,\Omega X] \to [A,\Omega X]$.

Remarks 3.11. (a) In the proof of Corollary 3.8 it was shown that $[A,\Omega X]$ is an S-group by considering the exact sequence

$$(*) \quad [A,K(\pi_n,n)] \to [A,\Omega X^{n+1}] \to [A,\Omega X^n].$$

However, a theorem of G.W. Whitehead (see [Berstein-Ganea 1961]) asserts that $[A,\Omega X]$ is actually a nilpotent group. Which special properties of the sequence (*) ensure that the nilpotency of $[A,K(\pi_n,n)]$ and $[A,\Omega X^n]$ imply that of $[A,\Omega X^{n+1}]$? (An extension of one nilpotent group by another is not nilpotent in general.)

(b) We wish to point out again that, although the results of this section are stated for the case when X has only a finite number of nontrivial homotopy groups, they hold for any space X provided A and B are finite-dimensional CW-complexes. Furthermore, the results hold if the loop space ΩX is replaced by any H-space that admits a classifying space (such as a topological group). Finally we note that our arguments are valid in the dual situation. That is, analogous expressions can be obtained for the rank of the kernel, image and fixed point group connected with homomorphisms $f_*: [\Sigma X, A] \to [\Sigma X, B]$ induced by maps $f: A \to B$. The proofs utilize homology decompositions instead of homotopy (Postnikov) decompositions.

4. Commutativity and Homotopy-Commutativity

The aim of this section is to show, by means of the methods developed in Chapter 3, that for an associative H-space G there exist maps $G \times G \to G$ satisfying certain commutativity conditions (Theorem 4.5). As will be explained in Remarks 4.6 this result is related to the work of other authors on homotopy-commutativity.

For technical reasons and purposes of illustration we work in a slightly more general setting and consider maps $A \times A \to \Omega Y$ instead of just dealing with maps $G \times G \to G$.

Let $\theta : A \times A \to A \times A$ be the switching map defined by $\theta(x,y) = (y,x)$. We denote by $\hat{A}$ the symmetric square of A, i.e., the orbit space of $A \times A$ under the action of θ, and by $q: A \times A \to \hat{A}$ the natural projection. Define $i: A \to \hat{A}$ by $i = qi_k : A \to A \times A \to \hat{A}$, where i_k imbeds A as the k-th factor of $A \times A$ for $k = 1,2$. For any space Y the maps q, i, i_k induce homomorphisms q^*, i^*, i_k^*:

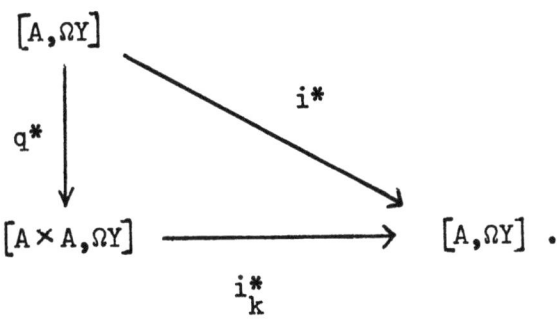

We note at this point that everything which is proved below for θ acting on $A \times A$ can equally well be done for a transitive permutation group of degree r acting on the r-fold cartesian product of A by permuting the factors.

Definition 4.1. An element $\mu \in [A \times A, \Omega Y]$ is said to be **commutative** $\mu \in \text{Im } q^*$. An element $\mu \in [A \times A, \Omega Y]$ is said to be **homotopy-commutative** if $\theta^* \mu = \mu$, where $\theta^* : [A \times A, \Omega Y] \to [A \times A, \Omega Y]$ is induced by θ. A commutative or homotopy-commutative element μ is said to be <u>of type α</u>, for α an element of $[A,\Omega Y]$, if $i_k^*\mu = \alpha$, $k = 1,2$.

Clearly any commutative element is homotopy-commutative. Proposition 4.2 below shows, however, that an appropriate multiple of a homotopy-commutative element is commutative.

<u>In Propositions 4.2, 4.3 and 4.4 we let A be a finite CW-complex and Y any 1-connected space.</u> Then $A \times A$ and A are also finite CW-complexes, and it follows from Lemma 3.1 that

$$(*) \begin{cases} [A \times A, \Omega Y] \cong [A \times A, \Omega Y^{(M)}], \\ [A, \Omega Y] \cong [A, \Omega Y^{(M)}] \end{cases}$$

for some Postnikov section $Y^{(M)}$ of Y. The isomorphisms (*) enable us to apply the results of Chapter 3 here. This will be done without explicit mention.

<u>Proposition 4.2.</u> If $\mu \in [A \times A, \Omega Y]$ is homotopy-commutative, then there exists an integer $N > 0$ such that the element $N\mu = \mu + \ldots + \mu \in [A \times A, \Omega Y]$ is commutative.

<u>Proof.</u> Let Γ be the group consisting of Θ and the identity operating on $A \times A$. By Proposition 3.10 $\rho([A \times A, \Omega Y]^\Gamma) = \rho(\text{Im} q^*)$. Then it follows by Proposition 2.5 (b) that the inclusion $\text{Im} q^* \subseteq [A \times A, \Omega Y]$ is an F-epimorphism. Thus there exists an N such that the N-fold multiple of μ is commutative.

Now we turn to the question of the existence of commutative and homotopy-commutative element of a given type.

<u>Proposition 4.3.</u> For any $\alpha \in [A,\Omega Y]$ there exist <u>positive</u> integers $N_c(\alpha)$ and $N_{hc}(\alpha)$ such that

(1) There is a commutative element $\mu \in [A \times A, \Omega Y]$ of type $N\alpha$ if and only if the integer N is a multiple of $N_c(\alpha)$.

(2) There is a homotopy-commutative element $\mu \in [A \times A, \Omega Y]$ of type $N\alpha$ if and only if the integer N is a multiple of $N_{hc}(\alpha)$.

Clearly $N_{hc}(\alpha)$ divides $N_c(\alpha)$.

Proof. Let $\xi_i \in [A \times A, \Omega Y]$ be commutative of type $N_i\alpha$, $i = 1,2$. Then $\xi_1 \pm \xi_2$ is commutative of type $(N_1 \pm N_2)\alpha$. Furthermore for any integer m clearly $m\xi_1$ is commutative of type $(mN_1)\alpha$. Thus all integers N such that there exists a commutative element of type $N\alpha$ form an ideal of the ring of integers, and are therefore multiples of an integer $N_c(\alpha)$. For the same reasons all integers N such that there exists a homotopy-commutative element of type $N\alpha$ are multiples of an integer $N_{hc}(\alpha)$. Since a commutative element is homotopy-commutative $N_{hc}(\alpha)$ divides $N_c(\alpha)$. Thus it suffices to show that $N_c(\alpha) \neq 0$, i.e., that there exists a commutative element of type $N\alpha$ for some $N \neq 0$. Now it is a result of Liao [1954, p. 526] that $i^m : H^m(A;\pi) \to H^m(A;\pi)$ is an epimorphism for all m and any coefficient group π. Therefore $i^* : [A, \Omega Y] \to [A, \Omega Y]$ is an F-epimorphism by Corollary 3.7. Thus for any $\alpha \in [A, \Omega Y]$ there exists a positive integer N such that $N\alpha = i^*\beta$ for some $\beta \in [A, \Omega Y]$. Then $\mu = q^*\beta$ is the desired commutative element of type $N\alpha$.

Once the existence of commutative elements has been established one may ask how many different commutative elements of a fixed type exist. In the following Proposition we give an easily applicable criterion for the existence of an <u>infinite</u> number of such elements.

Proposition 4.4. Consider a fixed $\alpha \in [A, \Omega Y]$, and let $N_c(\alpha)$, $N_{hc}(\alpha)$ be the integers of Proposition 4.3. The following three assertions are equivalent:

(1) For any N which is a multiple of $N_c(\alpha)$ there exists an infinite number of commutative elements $\mu \in [A \times A, \Omega Y]$ of type $N\alpha$.

(2) For any N which is a multiple of $N_{hc}(\alpha)$ there exists an infinite number of homotopy-commutative elements $\mu \in [A \times A, \Omega Y]$ of type $N\alpha$.

(3) For some m, $[\beta_m(\hat{A}) - \beta_m(A)]\rho(\pi_m(\Omega Y)) > 0$, where $\beta_m(X) = \rho(H_m(X))$.

Proof. Let i_o denote the restriction of i_k^* to the subgroup $[A \times A, \Omega Y]^\theta \subseteq [A \times A, \Omega Y]$. This subgroup contains Im q^*:

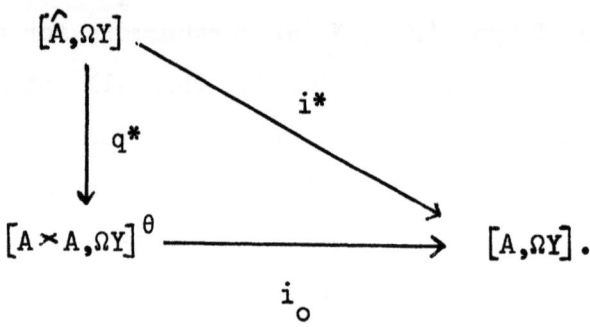

We conclude from Proposition 3.10 that q^* is an F-epimorphism. On the other hand it follows from III, Theorem 2.1 of [Borel, 1960] that $q^m : H^m(A;Q) \to H^m(A \times A;Q)$ is a monomorphism. Hence q^* is an F-monomorphism by Corollary 3.7. Thus q^* is an F-isomorphism and induces an F-isomorphism Ker $i^* \to$ Ker i_o. Both Ker i^* and Ker i_o are S-groups. Therefore by Lemma 2.6 the three statements

    (i) $q^*(\text{Ker } i^*)$ is infinite
   (ii)   Ker i_o is infinite
  (iii)   Ker i^* is infinite

are equivalent.

Now we observe that $q^* i^{*-1}(\beta)$ is the set of commutative elements of type $\beta \in [A, \Omega Y]$. Clearly $q^* i^{*-1}(\beta)$ is infinite if and only if the same holds for $q^*(\text{Ker } i^*)$. Thus assertion (1) of the Proposition is equivalent to statement (i) above. Similarly $i_o^{-1}(\beta)$ is the set of homotopy-commutative elements of type β. This set is infinite if and only if Ker i_o is infinite. Hence (2) and (ii) are equivalent. It remains to prove (3) $\Leftrightarrow$ (iii). It follows from [Liao, 1954, p. 526] quoted above that $\rho(\text{Ker } i^m) = \beta_m(\hat{A}) - \beta_m(A)$ where $i^m : H^m(\hat{A}) \to H^m(A)$ is induced by i. The equivalence (3) $\Leftrightarrow$ (iii) is now a consequence of Corollary 3.8, and the proof of the Proposition is complete.

Now let G be a connected associative H-space. By [Dold-Lashoff, 1959] there exists a space B_G such that G and ΩB_G are homotopy equivalent as H-spaces. We apply the preceding results on $[A \times A, \Omega Y]$ with $A = G$, $Y = B_G$ and $\alpha = \text{Id}_G$, the class of the identity map of G. An element $\mu \in [G \times G, G]$ will be called a <u>product of G</u>. A commutative or homotopy-commutative product of type $N \cdot \text{Id}_G \in [G, G]$, N an integer, is said to be <u>of type N</u>. The following Theorem is the main result of this Chapter.

<u>Theorem 4.5.</u> Let the finite CW-complex G be an associative H-space.
a) There exist positive integers $N_c(G)$ and $N_{hc}(G)$ such that
   (1) G admits a commutative product of type N if and only if N is a multiple of $N_c(G)$.
   (2) G admits a homotopy-commutative product of type N if and only if N is a multiple of $N_{hc}(G)$.

b) For any homotopy-commutative product μ of G there exists an integer $M > 0$ such that the M-fold multiple of μ is commutative.

c) The following three statements are equivalent:
   (1) For any N which is a multiple of $N_c(G)$ there exists an infinite number of different commutative products of type N.
   (2) For any N which is a multiple of $N_{hc}(G)$ there exists an infinite number of different homotopy-commutative products of type N.
   (3) For some m, $[\beta_m(\hat{G}) - \beta_m(G)]\rho(\pi_m(G)) > 0$.

<u>Proof.</u> Define $N_c(G)$ and $N_{hc}(G)$ to be the integers $N_c(\text{Id}_G)$ and $N_{hc}(\text{Id}_G)$ of Proposition 4.3; then a) follows. Proposition 4.2 implies b), and Proposition 4.4 implies c).

<u>Remarks 4.6.</u>
a) Using the formula of Richardson [1935] for the Betti numbers of $\hat{G}$ one shows easily that condition (3) of Theorem 4.5.c) is satisfied if and only if $\beta_m(G \# G)\rho(\pi_m(G)) > 0$ for some m ($G \# G = G \times G / G \vee G$). The latter inequality holds if and only if the group $[G \# G, G]$ is infinite (Corollary 3.4). But by Lemma 2 of [Arkowitz-Curjel, 1963$_{II}$] the group $[G \# G, G]$ is an 1-1-correspondence with the multiplications of G, i.e., with those products $\mu \in [G \times G, G]$ which restrict to Id_G on both factors

of $G \times G$. Thus we see that G admits an infinite number of commutative
and homotopy-commutative products of type $rN_c(G)$ and $sN_{hc}(G)$ for any
integers r and s if and only if G possesses an infinite number of
multiplications. The classical and exceptional Lie groups for which
this is the case are enumerated in Theorem 5 of [Arkowitz-Curjel, 1963_{II}].

b) Theorem 4.5 does for topological groups what results of James did
for spheres [1957], [1959_I]. James considers commutative and homotopy-
commutative products $S^n \times S^n \to S^n$ whose type is defined by means of the
suspension structure of S^n. He establishes the existence of integers
to which our N_c and N_{hc} are the analogues. It is amusing to note that
James' and our work meet in S^3 and S^1, the only spheres which are
topological groups.

c) We relate our terminology and notions to those of other authors.
A commutative element $\mu \in [G \times G, G]$ of type 1 is the homotopy class of
an <u>equivalent-commutative multiplication of G</u> in the sense of James
[1959_{II}]. Next let H be a topological group, $\mu_o \in [H \times H, H]$ the homotopy
class of the multiplication $H \times H \to H$ of H, G a subgroup of H and
$j : G \to H$ the inclusion homomorphism. The element $j\mu_o : G \times G \longrightarrow G \longrightarrow H$
is homotopy-commutative of type $j \in [G,H]$ if <u>G is homotopy-abelian in H</u>
in the sense of James-Thomas [1959]. To say that $N_{hc}(j) > 1$ means that
there is no homotopy-commutative element $\mu \in [G \times G, H]$ which restricts to
j on each of the factors of $G \times G$. Clearly $N_{hc}(j) > 1$ implies that G is
not homotopy-abelian in H, and in particular that G admits no homotopy-
commutative multiplication. In general, let $\phi : G \to H$ be an H-space
homomorphism of the associative H-spaces G and H. Then $N_c(G)$ and $N_c(H)$
are multiples of $N_c(\phi)$, and the same holds for $N_{hc}(G)$, $N_{hc}(H)$ and $N_{hc}(\phi)$.

d) If A is an orientable homology n-manifold then the <u>homology-type</u>
of a commutative or homotopy-commutative product $\mu \in [A \times A, A]$ can be
defined as the integer N by which the homomorphism $\mu_* i_{k*} : H_n(A) \longrightarrow H_n(A \times A) \longrightarrow H_n(A)$ multiplies a generator of $H_n(A)$. (This definition is
suggested in a recent paper of James-Thomas-Toda-Whitehead which treats
the case $A = S^n$.) By a result of Browder any H-space G which is a finite

CW-complex is an orientable homology n-manifold. It then follows from Theorem 4.5 by a homological argument that G admits commutative and homotopy-commutative products of homology-type N^r, where N is any multiple of $N_c(G)$ or $N_{hc}(G)$ and r is the dimension of the subspace of primitive elements of $H^*(G;Q)$.

e) One may ask for the numerical values (or at least estimates) of the invariants N_c and N_{hc}. In general it can be shown that for any $\alpha \in [A, \Omega Y]$ the integer $N_c(\alpha)$ divides $\prod_{s < \dim A} |k^s|$ where $|k^s|$ denotes the order of the s-th Postnikov invariant of ΩY. In the case $G = S^3$ James has shown that $N_c(S^3) = 4$ [1957] and $N_{hc}(S^3) = 2$ [1959$_I$]. There is the general result of Browder [1962] that a finite CW-complex X does not admit any homotopy-commutative multiplication if $H_*(X)$ has 2-torsion. Thus $N_{hc}(G) > 1$ and consequently $N_c(G) > 1$ if $H_*(G)$ has 2-torsion. Using this result, explicit computations for the groups SU(n), Sp(n) and the classification theorem for Lie groups, we have shown that $\underline{N_c(G) > 1}$ for any compact connected Lie group G other than a torus. James and Thomas have proved [1962] that a compact connected topological group G other than a torus is not homotopy-abelian. It is conceivable that such a space G nevertheless possesses a homotopy-commutative multiplication. However we make the

Conjecture. Let the finite CW-complex G be a homotopy-associative H-space not of the homotopy type of a torus. Then $N_{hc}(G) > 1$. A weaker form of this conjecture concerns compact connected Lie groups rather than homotopy-associative H-spaces. To prove the weaker conjecture it would suffice, by Browder's result and the classification theorem for Lie groups, to prove it for the groups SU(n) and Sp(n). By ad hoc methods the authors have succeeded in doing it for SU(3), SU(4) and SU(5).

5. The Rank of the Group of Homotopy Equivalences

In this chapter we completely determine the rank of the group of homotopy equivalences of a large class of spaces, namely those whose rational cohomology is an exterior algebra on odd dimensional generators. In addition to topological groups this class includes products of odd dimensional spheres, quaternionic, complex and some real Stiefel varieties and other homogeneous spaces. Aside from any intrinsic interest that the results in this section may have, they are included to illustrate how one can make statements on the rank of certain groups arising in homotopy theory whose group operation, being derived from composition of maps, is of a very different nature from those considered in the preceding chapter.

We adopt the following conventions: All group operations are written multiplicatively. The same symbol is used for a map $A \to B$ and its homotopy class in $[A,B]$. We let $\mathcal{E}(B)$ denote the group of homotopy classes of homotopy equivalences of the space B, where the group operation is defined by composition of maps. The subgroup of $\mathcal{E}(B)$ consisting of all elements which induce the identity automorphism on homotopy groups is written $\mathcal{E}_{\#}(B)$. First we consider two lemmas which deal with the group of equivalences of fairly general spaces. These lemmas, as well as a weaker form of Corollary 5.5, were stated without proof in our note [Arkowitz-Curjel, 1964].

Lemma 5.1. Let B be a 1-connected N-dimensional CW-complex and let $B^{(n)}$ be the n-th Postnikov section of B. Then

$$\mathcal{E}(B) \cong \mathcal{E}(B^{(N)}).$$

Proof. Since the projection $\phi : B \to B^{(N)}$ is an N-equivalence, $\phi_{\#} : [B,B] \to [B,B^{(N)}]$ is a bijection by Lemma 3.1. But it easily follows from known results on induced maps of Postnikov sections that $\phi^{*} : [B^{(N)},B^{(N)}] \to [B,B^{(N)}]$ is bijective. Thus by assigning to any map

$B \to B$ the induced map of N-th Postnikov sections $B^{(N)} \to B^{(N)}$, we obtain a one-to-one correspondence $[B,B] \to [B^{(N)}, B^{(N)}]$. Since this bijection is compatible with the composition operation, the Lemma is proved.

Clearly the homomorphism $\mathcal{E}(B^{(n)}) \to \mathcal{E}(B^{(n-1)})$ obtained by restricting a map to a lower Postnikov section gives rise to a homomorphism $\rho_n : \mathcal{E}_{\#}(B^{(n)}) \to \mathcal{E}_{\#}(B^{(n-1)})$. The next lemma gives some information on the kernel of ρ_n.

Lemma 5.2. Let $T^n(B)$ denote the kernel of the homomorphism $H^n(B; \pi_n(B)) \to \mathrm{Hom}(\pi_n(B), \pi_n(B))$. Then there exists a homomorphism $\theta_n : T^n(B) \to \mathcal{E}_{\#}(B^{(n)})$ such that the sequence

$$T^n(B) \xrightarrow{\theta_n} \mathcal{E}_{\#}(B^{(n)}) \xrightarrow{\rho_n} \mathcal{E}_{\#}(B^{(n-1)})$$

is exact.

We only sketch the proof. It is well known that there is an operation of the fibre $K = K(\pi_n(B), n)$ of the fibration $K \xrightarrow{i} B^{(n)} \xrightarrow{p} B^{(n-1)}$ on the total space $B^{(n)}$. This operation determines an operation of $[A,K]$ on $[A, B^{(n)}]$ for any space A, i.e., if $\alpha \in [A,K]$ and $\beta \in [A, B^{(n)}]$ we obtain $\beta^\alpha \in [A, B^{(n)}]$ (for properties of this operation see [Eckmann-Hilton, 1960]). Now we take $A = B^{(n)}$ and observe that if $\alpha \in T^n(B^{(n)}) \subseteq [B^{(n)}, K]$ then $\mathrm{Id}^\alpha \in \mathcal{E}_{\#}(B^{(n)}) \subseteq [B^{(n)}, B^{(n)}]$. Since $T^n(B^{(n)}) \cong T^n(B)$ we obtain a transformation $\theta_n : T^n(B) \to \mathcal{E}_{\#}(B^{(n)})$. A straightforward but rather long computation shows that θ_n is a homomorphism. Exactness of the sequence $T^n(B) \xrightarrow{\theta_n} \mathcal{E}_{\#}(B^{(n)}) \xrightarrow{\rho_n} \mathcal{E}_{\#}(B^{(n-1)})$ easily follows from standard properties of the operation.

For the remainder of this section Y shall always stand for a 1-connected N-dimensional CW-complex whose rational cohomology $H^*(Y;Q)$ is an exterior algebra on generators of odd degree n_i, $i = 1,\ldots,k$. The following two facts about the spaces Y shall be used. First of all, it is well known that the homotopy groups of Y are the same, modulo torsion, as the homotopy groups of the product of spheres $S^{n_1} \times \ldots \times S^{n_k}$. Secondly, all the Postnikov invariants of Y are cohomology elements of finite order [Thom, 1956].
For the spaces Y, Lemma 5.2 can be sharpened in the following way.

Proposition 5.3. In the exact sequence of Lemma 5.2

$$T^n(Y) \xrightarrow{\theta_n} \mathcal{E}_{\#}(Y^{(n)}) \xrightarrow{\rho_n} \mathcal{E}_{\#}(Y^{(n-1)}),$$

θ_n is an F-monomorphism and ρ_n is an F-epimorphism.

Proof. Any element $\alpha \in \mathcal{E}_{\#}(Y^{(n-1)})$ induces an automorphism α^* of the cohomology group $H^n = H^n(Y^{(n-1)}; \pi_{n+1}(Y))$ and, consequently, an automorphism α^*_T of the torsion subgroup H^n_T of H^n. Since the automorphism group of a finite group is finite, it follows that the composition $(\alpha^*_T)^r = \alpha^*_T \circ \ldots \circ \alpha^*_T = \mathrm{Id}$ for some integer r. But the Postnikov invariant $k^n \in H^n$ can be regarded as an element of H^n_T. Thus $(\alpha^*)^r(k^n) = (\alpha^*_T)^r(k^n) = k^n$, and so there is commutativity in the square

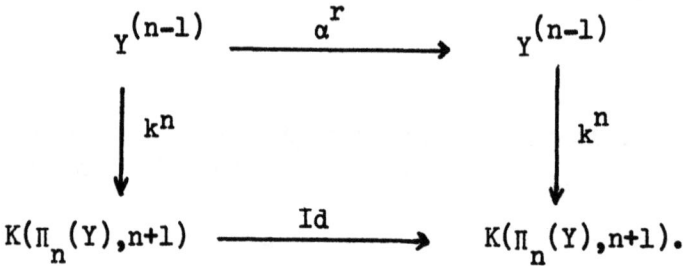

With $p : Y^{(n)} \to Y^{(n-1)}$ the fibre map there is then a map $\beta : Y^{(n)} \to Y^{(n)}$ such that $\alpha^r p = p\beta$ and $\beta_* = \mathrm{Id} : \pi_n(Y^{(n)}) \to \pi_n(Y^{(n)})$. Therefore β is in $\mathcal{E}_{\#}(Y^{(n)})$ and $\rho_n(\beta) = \alpha^r$. This proves that ρ_n is an F-epimorphism.

To show that θ_n is an F-monomorphism, choose $\alpha \in T^n(Y^{(n)})$ such that $\mathrm{Id}^\alpha = \mathrm{Id}$. We first investigate the effect of $\alpha : Y^{(n)} \to K$ on n-dimensional rational homology, where $K = K(\pi_n(Y), n)$. By definition Id^α is the composition

$$Y^{(n)} \xrightarrow{\Delta} Y^{(n)} \times Y^{(n)} \xrightarrow{\mathrm{Id} \times \alpha} Y^{(n)} \times K \xrightarrow{\nu} Y^{(n)}.$$

Here Δ is the diagonal map and ν is the operation of fibre on total space of the fibration $K \xrightarrow{i} Y^{(n)} \xrightarrow{p} Y^{(n-1)}$. If $x \in H_n(Y^{(n)}; \mathbb{Q})$ then in the group $\sum_{r+s=n} H_r(Y^{(n)}; \mathbb{Q}) \otimes H_s(Y^{(n)}; \mathbb{Q})$

$$\Delta_*(x) = x \otimes 1 + 1 \otimes x + \sum x_i' \otimes x_j''$$

where the x_j' are homology elements having positive degree $< n$. Since $H_r(K;Q) = 0$ for $0 < r < n$ we have that

$$(\text{Id} \otimes \alpha_*)\, \Delta_*(x) = x \otimes 1 + 1 \otimes \alpha_*(x)$$

and, consequently, using standard properties of the operation ν, that

$$\nu_*(\text{Id} \times \alpha)_* \Delta_*(x) = x + i_* \alpha_*(x).$$

Thus our hypothesis $\nu(\text{Id} \times \alpha)\Delta = \text{Id}$ implies

$$x + i_* \alpha_*(x) = x.$$

Hence we have proved

$$i_* \alpha_* = 0 : H_n(Y^{(n)};Q) \to H_n(Y^{(n)};Q).$$

Now the Theorem of [Serre, 1951 p. 468] applied to the fibration $K \xrightarrow{i} Y^{(n)} \xrightarrow{p} Y^{(n-1)}$ yields the diagram

$$H_{n+1}(Y^{(n-1)};Q) \longrightarrow H_n(K;Q) \xrightarrow{i_*} H_n(Y^{(n)};Q)$$

with k_*^n arrow down and isomorphism $\cong$ up to

$$H_{n+1}(K(\pi_n(Y), n+1); Q)$$

with exact row and commutative triangle. Since k^n is of finite order, $k_*^n = 0$. Thus $i_* : H_n(K;Q) \to H_n(Y^{(n)};Q)$ is a monomorphism. It then follows from $i_* \alpha_* = 0$ that

$$\alpha_* = 0 : H_n(Y^{(n)};Q) \longrightarrow H_n(K;Q).$$

Therefore $\alpha^* = 0 : H^n(K;Q) \to H^n(Y^{(n)};Q)$. By Corollary 3.6, α has finite order. Consequently we have shown that α is an element of finite order if $\theta_n(\alpha) = \text{Id}^\alpha = \text{Id}$. Thus θ_n is an F-monomorphism. This completes the proof of Proposition 5.3.

Next we state a Proposition whose proof we omit because it is based largely on arguments not pertaining to the subject of these notes. (Actually, our result gives more detailed information than is stated below.)

<u>Proposition.</u> If $\rho(\pi_n(Y)) > 1$ for some $n \leq N$, then $\mathcal{E}(Y)$ contains a free subgroup on at least two generators.

We next prove the main theorem of this chapter.

<u>Theorem 5.4.</u> Let Y be a 1-connected N-dimensional CW-complex whose rational cohomology is an exterior algebra on generators of odd dimension n_i, $i = 1,\ldots,k$. Then $\mathcal{E}(Y)$ has finite rank if and only if all the n_i are distinct. In this case,

$$\rho(\mathcal{E}(Y)) = \sum_{i=1}^{k} (\beta_{n_i}(Y)-1)$$

where $\beta_{n_i}(Y)$ denotes the n_i-th Betti number of Y.

<u>Proof.</u> Clearly all the n_i are distinct if and only if the rank of $\pi_n \leq 1$ for all $n \leq N$, where $\pi_n = \pi_n(Y)$. Thus if $n_i = n_j$ for $i \neq j$, then $\mathcal{E}(Y)$ contains a free subgroup on at least two generators by the preceding Proposition. Hence $\mathcal{E}(Y)$ is not of finite rank (see Remark 2.7(c)).

Now we suppose $\rho(\pi_n) \leq 1$ for all $n \leq N$. Consider the exact sequence

$$(*) \quad 1 \to \mathcal{E}_\#(Y^{(N)}) \xrightarrow{j} \mathcal{E}(Y^{(N)}) \xrightarrow{J} \sum_{n \leq N} \text{Aut } \pi_n$$

where j is the inclusion and J is the homomorphism which assigns to each element of $\mathcal{E}(Y^{(N)})$ its induced homotopy automorphism. Clearly the hypothesis $\rho(\pi_n) \leq 1$ guarantees that $\sum_{n < N} \text{Aut } \pi_n$ is a finite group. Since $\mathcal{E}(Y) \cong \mathcal{E}(Y^{(N)})$ by Lemma 5.1, we have

$$\rho(\mathcal{E}(Y)) = \rho(\mathcal{E}_{\#}(Y)).$$

But by Proposition 5.3 the sequence $T^n(Y) \xrightarrow{\theta_n} \mathcal{E}_{\#}(Y^{(n)}) \xrightarrow{\rho_n} \mathcal{E}_{\#}(Y^{(n-1)})$ satisfies the hypotheses of Lemma 2.8. Thus

$$\rho(\mathcal{E}_{\#}(Y^{(n)})) = \rho(\mathcal{E}_{\#}(Y^{(n-1)})) + \rho(T^n(Y)).$$

Now $\rho(T^n(Y))$ is easily determined: $T^n(Y)$ is the kernel of $h'_n\eta$

$$H^n(Y;\pi_n) \xrightarrow{\eta} \mathrm{Hom}(H_n(Y),\pi_n) \xrightarrow{h'_n} \mathrm{Hom}(\pi_n,\pi_n)$$

where $\pi_n = \pi_n(Y)$, η is the homomorphism of the universal coefficient theorem for cohomology and h'_n is the homomorphism induced by the Hurewicz homomorphism $h_n: \pi_n \to H_n(Y)$. Since η is an F-isomorphism — it is an epimorphism whose kernel is a finite group — it suffices to determine $\rho(\mathrm{Ker}\ h'_n)$. Clearly $\mathrm{Ker}\ h'_n \cong \mathrm{Hom}(\mathrm{coker}\ h_n, \pi_n)$, and so $\rho(\mathrm{Ker}\ h'_n) = \rho(\mathrm{coker}\ h_n) \cdot \rho(\pi_n)$. But a result of [Cartan-Serre, 1952] asserts that, under the hypothesis on the rational cohomology of Y, the kernel of h_n is a finite group. Thus

$$\rho(\mathrm{coker}\ h_n) = \beta_n(Y) - \rho(\mathrm{Im}\ h_n)$$

$$= \beta_n(Y) - \rho(\pi_n).$$

Therefore

$$\rho(T^n(Y)) = \rho(\mathrm{coker}\ h_n) \cdot \rho(\pi_n)$$

$$= (\beta_n(Y) - \rho(\pi_n)) \cdot \rho(\pi_n)$$

$$= \begin{cases} \beta_{n_i}(Y) - 1 & n = n_i \\ 0 & n \neq n_i,\ i=1,\ldots k. \end{cases}$$

Putting things together, we have

$$\rho(\mathcal{E}(Y)) = \rho(\mathcal{E}_{\#}(Y^{(N)}))$$

$$= \sum_{n=2}^{N} \rho(T^n(Y))$$

$$= \sum_{i=1}^{k} (\beta_{n_i}(Y) - 1).$$

The proof is now complete.

Corollary 5.5. Let Y satisfy the hypotheses of Theorem 5.4. Then $\mathcal{E}(Y)$ is finite if and only if $\beta_{n_i}(Y) = 1$ for all $i = 1,\ldots,k$.

Proof. If $\mathcal{E}(Y)$ is finite then plainly $\mathcal{E}(Y)$ has rank zero, and so $\beta_{n_i}(Y) = 1$ by Theorem 5.4. If on the other hand all $\beta_{n_i}(Y) = 1$, then all the n_i are distinct. Thus $\rho(\pi_n) \leq 1$ for all $n \leq N$, and so $\text{Aut}\pi_n$ is a finite group. It then follows from the sequence (*) of the preceding proof that $\mathcal{E}(Y)$ is finite if $\mathcal{E}_{\#}(Y^{(N)})$ is. But by Lemma 5.2 $\mathcal{E}_{\#}(Y^{(N)})$ is an S-group (see Chapter 2). From Remark 2.6(b) we deduce that $\mathcal{E}_{\#}(Y^{(N)})$ is finite if its rank is zero. However,

$$\rho(\mathcal{E}_{\#}(Y^{(N)})) = \rho(\mathcal{E}(Y))$$

$$= \sum_{i=1}^{k} (\beta_{n_i}(Y) - 1)$$

$$= 0.$$

Thus $\mathcal{E}_{\#}(Y^{(N)})$ and consequently $\mathcal{E}(Y)$ are finite groups. This establishes the Corollary.

Remark 5.6. Since the rational cohomology algebra of Lie groups and many homogeneous spaces have been computed (see [Borel, 1953]), Theorem 5.4 enables us to determine the rank of the group of equivalences of these spaces. Thus, for instance, $\mathcal{E}(\text{Spin}(4r))$ does not have finite rank, but $\mathcal{E}(\text{Spin}(n))$ does for $n \neq 4r$. Furthermore, all 1-connected representatives of the exceptional Lie structure classes other than E_6 have a finite group of homotopy equivalences, while $\rho(\mathcal{E}(E_6)) = 1$.

As a final illustrative example consider the special unity group SU_n. If $n \leq 7$ then $\mathcal{E}(SU_n)$ is finite. For larger values of n we have for instance

$$\rho(\mathcal{E}(SU_8)) = 1, \quad \rho(\mathcal{E}(SU_{11})) = 7, \quad \rho(\mathcal{E}(SU_{15})) = 31.$$

In general

$$\rho(\mathcal{E}(SU_n)) = \rho(\mathcal{E}(SU_{n-1})) + \theta(2n-1)$$

where $\theta(2n-1)$ is the number of distinct ways of writing $2n-1$ as a sum of the integers $3, 5, 7, \ldots, 2n-3$.

Bibliography

M. Arkowitz and C.R. Curjel

1963_I: Homotopy commutators of finite order (I), Quart.J.Math. Oxford (2), $\underline{14}$ (1963), 213-219.

1963_{II}: On the number of multiplications of an H-space, Topology $\underline{2}$ (1963), 205-208.

1964: The group of homotopy equivalences of a space, Bull.Amer. Math.Soc. $\underline{70}$ (1964), 293-296.

I. Berstein and T. Ganea, Homotopical nilpotency, Illinois J.Math. $\underline{5}$ (1961), 99-130.

A. Borel

1953: Sur la cohomologie des espaces fibrés principaux et des espaces homogènes de groupes de Lie compacts, Ann. of Math. $\underline{57}$ (1953), 115-207.

1960: Seminar on transformation groups, Ann. of Math.Studies No. 46, Princeton University Press, Princeton, N.J., 1960.

W. Browder, Homotopy commutative H-spaces, Ann. of Math. $\underline{75}$ (1962), 283-311.

H. Cartan and J.-P. Serre, Espaces fibrés et groupes d'homotopie, II. Applications. C.R.Acad.Sci.Paris $\underline{234}$ (1952), 393-395.

A. Dold and R. Lashof, Principal quasifibrations and fibre homotopy equivalence of bundles, Illinois J.Math. $\underline{3}$ (1959), 285-305.

M. Hall Jr., The theory of groups, Macmillan, New York, N.Y., 1959.

K.A. Hirsch, On infinite soluble groups, I, Proc.London Math.Soc. (2) $\underline{44}$ (1938), 53-60.

I. James

1957: Commutative products on spheres, Cambridge Philosophical Soc. $\underline{53}$ (1957), 63-68.

1959_I: Products on spheres, Mathematika $\underline{6}$ (1959), 1-13.

1959_{II}: The ten types of H-spaces (mimeographed), 1959.

I. James and E. Thomas
- 1959: Which Lie groups are homotopy-abelian? Proc.Nat.Acad.Sci. USA **45** (1959), 737-740.

- 1962: Homotopy-abelian topological groups, Topology **1** (1962), 237-240.

S.A. Jennings, The group ring of a class of infinite nilpotent groups, Canadian J.Math. **7** (1955), 169-187.

S.D. Liao, On the topology of cyclic products of spheres, Trans.Amer. Math.Soc. **77** (1954), 520-551.

D. Puppe, Homotopiemengen und ihre induzierten Abbildungen I, Math. Zeitschrift **69** (1958), 299-344.

M. Richardson, On the homology characters of symmetric products, Duke Math.J. **1** (1935), 50-69.

J.-P. Serre, Homologie singulière des espaces fibrés, Ann. of Math. **54** (1951), 425-505.

R. Thom, L'homologie des espaces fonctionnels, Colloque de Topologie Algèbrique, pp. 29-39, Louvain, 1956.

J.H.C. Whitehead, Combinatorial homotopy, I, Bull Amer.Math.Soc. **55** (1949), 213-245.

H. Zassenhaus, The theory of groups, 2nd edition, Chelsea, New York, N.Y., 1958.

Lecture Notes in Mathematics

Bisher erschienen/Already published

Vol. 1: J. Wermer, Seminar über Funktionen-Algebren,
IV, 30 Seiten. 1964. DM 3,80

Vol. 2: A. Borel, Cohomologie des espaces localement
compacts d'après J. Leray.
IV, 93 pages. 1964. DM 9,–

Vol. 3: J. F. Adams, Stable Homotopy Theory.
2nd. revised edition. IV, 78 pages. 1966. DM 7,80

Vol. 5: J.-P. Serre, Cohomologie Galoisienne.
Troisième édition. VIII, 214 pages. 1965. DM 18,–

Vol. 6: H. Hermes, Eine Termlogik mit Auswahloperator.
IV, 42 Seiten. 1965. DM 5,80

Vol. 7: Ph. Tondeur, Introduction to Lie Groups
and Transformation Groups.
VIII, 176 pages. 1965. DM 13,50

Vol. 8: G. Fichera, Linear Elliptic Differential
Systems and Eigenvalue Problems.
IV, 176 pages. 1965. DM 13.50

Vol. 9: P. L. Ivănescu, Pseudo-Boolean Programming and
Applications. IV, 50 pages. 1965. DM 4,80

Vol. 10: H. Lüneburg, Die Suzukigruppen und ihre
Geometrien. VI, 111 Seiten. 1965. DM 8,–

Vol. 11: J.-P. Serre, Algèbre Locale. Multiplicités.
Rédigé par P. Gabriel. Seconde édition.
VIII, 192 pages. 1965. DM 12,–

Vol. 12: A. Dold, Halbexakte Homotopiefunktoren.
II, 157 Seiten. 1966. DM 12,–

Vol. 13: E. Thomas, Seminar on Fiber Spaces.
VI, 45 pages. 1966. DM 4,80

Vol. 14: H. Werner, Vorlesung über Approximations-
theorie. IV, 184 Seiten und 12 Seiten Anhang.
1966. DM 14,–

Vol. 15: F. Oort, Commutative Group Schemes.
VI, 133 pages. 1966. DM 9,80

Vol. 16: J. Pfanzagl and W. Pierlo, Compact Systems
of Sets. IV, 48 pages. 1966. DM 5,80

Vol. 17: C. Müller, Spherical Harmonics.
IV, 46 pages. 1966. DM 5,–

Vol. 18: H.-B. Brinkmann, und D. Puppe,
Kategorien und Funktoren.
XII, 107 Seiten. 1966. DM 8,–

Vol. 19: G. Stolzenberg, Volumes, Limits and Extensions
of Analytic Varieties. IV, 45 pages. 1966. DM 5,40

Vol. 20: R. Hartshorne, Residues and Duality.
VIII, 423 pages. 1966. DM 20,–

Vol. 21: Seminar on Complex Multiplication. By A. Borel,
S. Chowla, C. S. Herz, K. Iwasawa, J.-P. Serre.
IV, 102 pages. 1966. DM 8,–

Vol. 22: H. Bauer, Harmonische Räume und ihre Potential-
theorie. IV, 175 Seiten. 1966. DM 14,–

Vol. 23: P. L. Ivănescu and S. Rudeanu, Pseudo-Boolean
Methods for Bivalent Programming.
120 pages. 1966. DM 10,–

Vol. 24: J. Lambek, Completions of Categories. IV, 69 pages.
1966. DM 6,80

Vol. 25: R. Narasimhan, Introduction to the Theory of
Analytic Spaces. IV, 143 pages. 1966. DM 10,–

Vol. 26: P.-A. Meyer, Processus de Markov. IV, 190 pages.
1967. DM 15,–

Vol. 27: H. P. Künzi und S. T. Tan, Lineare Optimierung
großer Systeme. VI, 124 Seiten. 1966. DM 12,–

Vol. 28: P. E. Conner and E. E. Floyd, The Relation of
Cobordism to K-Theories. VIII, 113 pages.

Vol. 29: K. Chandrasekharan, Einführung in die
Analytische Zahlentheorie. VI, 199 Seiten.
1966. DM 16,80

Vol. 30: A. Frölicher and W. Bucher, Calculus in Vector
Spaces without Norm. XII, 146 pages. 1966. DM 12,–

Vol. 31: Probability Methods in Analysis. Chairman:
D. A. Kappos. IV, 329 pages. 1967. DM 20,–

Beschaffenheit der Manuskripte

Die Manuskripte werden photomechanisch vervielfältigt; sie müssen daher in sauberer Schreibmaschinenschrift geschrieben sein. Handschriftliche Formeln bitte nur mit schwarzer Tusche oder roter Tinte eintragen. Korrekturwünsche werden in der gleichen Maschinenschrift auf einem besonderem Blatt erbeten (Zuordnung der Korrekturen im Text und auf dem Blatt sind durch Bleistiftziffern zu kennzeichnen). Der Verlag sorgt dann für das ordnungsgemäße Tektieren der Korrekturen. Falls das Manuskript oder Teile desselben neu geschrieben werden müssen, ist der Verlag bereit, dem Autor bei Erscheinen seines Bandes einen angemessenen Betrag zu zahlen. Die Autoren erhalten 25 Freiexemplare.

Manuskripte, in englischer, deutscher oder französischer Sprache abgefaßt, nimmt Prof. Dr. A. Dold, Mathematisches Institut der Universität Heidelberg, Tiergartenstraße oder Prof. Dr. B. Eckmann, Eidgenössische Technische Hochschule, Forschungsinstitut für Mathematik, Zürich, entgegen.

Cette série a pour but de donner des informations rapides, de niveau élevé, sur des développements récents en mathématiques, aussi bien dans la recherche que dans l'enseignement supérieur. On prévoit de publier

1. des versions préliminaires de travaux originaux et de monographies

2. des cours spéciaux portant sur un domaine nouveau ou sur des aspects nouveaux de domaines classiques

3. des rapports de séminaires

4. des conférences faites à des congrès ou des colloquiums

En outre il est prévu de publier dans cette série, si la demande le justifie, des rapports de séminaires et des cours multicopiés ailleurs qui sont épuisés.

Dans l'intérêt d'une grande actualité les contributions pourront souvent être d'un caractère provisoire; le cas échéant, les démonstrations ne seront données qu'en grande ligne, et les résultats et méthodes pourront également paraître ailleurs. Par cette série de »prépublications« les éditeurs Springer espèrent rendre d'appréciables services aux instituts de mathématiques par le fait qu'une réserve suffisante d'exemplaires sera toujours à disposition et que les intéressés pourront plus facilement être atteints. Les annonces dans les revues spécialisées, les inscriptions aux catalogues et les copyrights faciliteront pour les bibliothèques mathématiques la tâche de dresser une documentation complète.

Présentation des manuscrits

Les manuscrits étant reproduits par procédé photomécanique, doivent être soigneusement dactylographiés. Il est demandé d'écrire à l'encre de Chine ou à l'encre rouge les formules non dactylographiées. Des corrections peuvent également être dactylographiées sur une feuille séparée (prière d'indiquer au crayon leur ordre de classement dans le texte et sur la feuille), la maison d'édition se chargeant ensuite de les insérer à leur place dans le texte. S'il s'avère nécessaire d'écrire de nouveau le manuscrit, soit complètement, soit en partie, la maison d'édition se déclare prête à se charger des frais à la parution du volume. Les auteurs recoivent 25 exemplaires gratuits.

Les manuscrits en anglais, allemand ou français peuvent être adressés au Prof. Dr. A. Dold, Mathematisches Institut der Universität Heidelberg, Tiergartenstraße ou Prof. Dr. B. Eckmann, Eidgenössische Technische Hochschule, Forschungsinstitut für Mathematik, Zürich.

If you have any concerns about our products,
you can contact us on
ProductSafety@springernature.com

In case Publisher is established outside the EU,
the EU authorized representative is:
**Springer Nature Customer Service Center GmbH
Europaplatz 3, 69115 Heidelberg, Germany**

Printed by Libri Plureos GmbH
in Hamburg, Germany